PATTERNS OF LIFE

Finding Shelter

Daphne Butler

SIMON & SCHUSTER
YOUNG BOOKS

This book was conceived for
Simon & Schuster Young Books by
Globe Enterprises of Nantwich, Cheshire

Design: SPL Design
Photographs: Zefa except for
NHPA (cover, title page, 10, 12, 15tl)

First published in Great Britain in 1993
by Simon & Schuster Young Books
Campus 400, Maylands Avenue
Hemel Hempstead, Herts HP2 7EZ

Printed and bound in Singapore
by Kim Hup Lee Printing Co Pte Ltd

A catalogue record for this book is available
from the British Library
ISBN 0 7500 1291 9

Contents

Whatever the weather

What's the weather like where you live? Is it cold sometimes, or rainy, or perhaps very hot?

Almost all creatures need somewhere to shelter from the weather for some of the time in almost every part of the world.

Warm coats

Different animals have different ways of living with the weather.

Some grow a thick hairy coat in autumn. This traps air keeping their skin warm and dry.

In spring, the extra hair falls out leaving only a thin coat for the hot summer.

Sleeping until spring

In autumn, some animals find a safe place to sleep. They stay there, hibernating till spring when the weather warms once again.

Fish lie still at the bottom of lakes and rivers. In the cold water under the ice, they wait for the thaw.

Shady places

Animals living in hot countries don't need thick coats. They need shady places to protect them from the heat of the sun.

Some climb high in the leafy branches. Here they might also find a cooling breeze.

Safe places

Animals also need safe places in which to hide from their enemies and where their babies can stay until they have grown strong.

Holes in the ground

Some animals dig themselves holes in the ground. Underground it is safe, and also cool in summer.

In winter, a hole is a safe place to hibernate away from the weather.

Nest building

Some animals build nests.

Birds collect twigs, grass or moss. Wasps and hornets make theirs with walls like paper.

How do you think spiders build their nests?

What sort of animals make strange-looking hills?

Building houses

People also build nests—they build houses with strong walls and roofs. Their houses are warm and dry in winter and cool and shady in summer.

Living with people

Some animals move in with people.

The pictures show some examples.
Can you name the animals?

Why do you think some animals
want to live with people?

Farm animals

Many farmers move their animals close to their farms in winter. They may even keep them inside barns or sheds.

Why do you think farmers do this?

Living anywhere

All animals, including people, learn how to live in their surroundings.

But people can travel to any part of the world. They can change their clothes and their houses to suit the weather.

They are very successful animals.

Useful words

birds Warm-blooded animals with feathers that they can fluff out to keep them warm in cold weather. The feathers are also waterproof.

environment Surroundings that affect the well-being of an animal such as its food supply and the weather conditions.

hibernating A way of sleeping through cold weather. Hibernating animals 'close down'—their body temperatures drop and their heart beats slow down.

mammals Warm-blooded animals usually having a coat of waterproof hair or fur. Air becomes trapped in the hair keeping them warm. Humans have little body hair but can make warm clothes for themselves.

nests Safe places made by some animals for their babies. Humans build houses in which to live and have their children.

predators Animals that prey on other animals for their food.

Index